Highlights of GAO-10-331, a report to the Subcommittee on Air and Land Forces, Committee on Armed Services, House of Representatives

March 2010

UNMANNED AIRCRAFT SYSTEMS

Comprehensive Planning and a Results-Oriented Training Strategy Are Needed to Support Growing Inventories

Why GAO Did This Study

The Department of Defense (DOD) requested about $6.1 billion in fiscal year 2010 for new unmanned aircraft systems (UAS) and for expanded capabilities in existing ones. To support ongoing operations, the Air Force and Army have acquired a greater number of larger systems. GAO was asked to determine the extent to which (1) plans were in place to account for the personnel, facilities, and communications infrastructure needed to support Air Force and Army UAS inventories; (2) DOD addressed challenges that affect the ability of the Air Force and the Army to train personnel for UAS operations; and (3) DOD updated its publications that articulate doctrine and tactics, techniques, and procedures to reflect the knowledge gained from using UAS in ongoing operations. Focusing on UAS programs supporting ongoing operations, GAO reviewed the services' program and funding plans in light of DOD's requirements definition and acquisition policy; interviewed UAS personnel in the United States and in Iraq about training experiences; and reviewed joint, multiservice, and service-specific publications.

What GAO Recommends

GAO recommends, among other things, that DOD conduct comprehensive planning as part of the decision-making process to field new systems or expand existing capabilities and that DOD develop a results-oriented strategy for addressing training challenges. DOD generally agreed with the recommendations.

View GAO-10-331 or key components.
For more information, contact Sharon Pickup at (202) 512-9619 or pickups@gao.gov.

What GAO Found

DOD continues to increase UAS inventories, but in some cases, the Air Force and the Army lack robust plans that account for the personnel, facilities, and some communications infrastructure to support them. Regarding personnel, the Air Force and the Army have identified limitations in their approaches to provide personnel to meet current and projected UAS force levels, but they have not yet fully developed plans to supply needed personnel. Further, although DOD has recently requested funding and plans to request additional funds, the Air Force and the Army have not completed analyses to specify the number and type of facilities needed to support UAS training and operations. Having identified a vulnerability to the communications infrastructure network used to control UAS missions, the Air Force is taking steps to mitigate the risk posed by a natural or man-made disruption to the network but has not formalized a plan in the near term to provide for the continuity of UAS operations in the event of a disruption. While DOD guidance encourages planning for factors needed to operate and sustain a weapon system program in the long term, several factors have contributed to a lag in planning efforts, such as the rapid fielding of new systems and the expansion of existing ones. In the absence of comprehensive planning, DOD does not have reasonable assurance that Air Force and Army approaches will support current and projected UAS inventories. The lack of comprehensive plans also limits the ability of decision makers to make informed funding choices.

DOD has not developed a results-oriented strategy to resolve challenges that affect the ability of the Air Force and the Army to train personnel for UAS operations. GAO found that the limited amount of DOD-managed airspace adversely affected the amount of training that personnel conducted to prepare for deployments. As UAS are fielded in greater numbers, DOD will require access to more airspace for training; for example, DOD estimated that based on planned UAS inventories in fiscal year 2013, the military services will require more than 1 million flight hours to train UAS personnel within the United States. Further, Air Force UAS personnel and Army ground units have limited opportunities to train together in a joint environment, and they have not maximized the use of available assets during training. Current UAS simulators also have limited capabilities to enhance training. DOD has commenced initiatives to address training challenges, but it has not developed a results-oriented strategy to prioritize and synchronize these efforts. Absent a strategy, DOD will not have a sound basis for prioritizing resources, and it cannot be assured that the initiatives will address limitations in Air Force and Army training approaches.

In many cases, DOD's UAS publications articulating doctrine and tactics, techniques, and procedures did not include updated information needed by manned and unmanned aircraft operators, military planners, and ground units to understand current practices and capabilities. Such information can serve as the foundation for effective joint training programs and can assist military personnel in integrating UAS on the battlefield.

United States Government Accountability Office

Contents

Letter		1
	Background	4
	Plans Are Not in Place to Fully Account for the Personnel, Facilities, and Some Communications Infrastructure Needed to Support Air Force and Army UAS Programs	10
	DOD Has Not Resolved Challenges That Affect the Ability of the Air Force and the Army to Train Personnel for UAS Operations	22
	DOD Has Not Fully Incorporated Knowledge Gained from Ongoing UAS Operations in Key Publications	32
	Conclusions	37
	Recommendations for Executive Action	37
	Agency Comments and Our Evaluation	39
Appendix I	**Scope and Methodology**	43
Appendix II	**Comments from the Department of Defense**	47
Appendix III	**GAO Contact and Staff Acknowledgments**	51
Tables		
	Table 1: Military Services' Inventories of Selected Unmanned Aircraft	6
	Table 2: DOD's Budget Requests for UAS (Fiscal Years 2007 through 2010)	7
	Table 3: DOD Organizations and Initiatives Addressing UAS Training Challenges	30
Figures		
	Figure 1: Line-of-Sight UAS Operational Concept	17
	Figure 2: Beyond-the-Line-of-Sight UAS Operational Concept	19

Abbreviations

DOD   Department of Defense
ERMP   Extended Range Multi-Purpose
UAS   unmanned aircraft systems

This is a work of the U.S. government and is not subject to copyright protection in the United States. The published product may be reproduced and distributed in its entirety without further permission from GAO. However, because this work may contain copyrighted images or other material, permission from the copyright holder may be necessary if you wish to reproduce this material separately.

United States Government Accountability Office
Washington, DC 20548

March 26, 2010

The Honorable Adam Smith
Chairman
The Honorable Roscoe Bartlett
Ranking Member
Subcommittee on Air and Land Forces
Committee on Armed Services
House of Representatives

Battlefield commanders have experienced a high level of mission success in ongoing operations with capabilities provided by unmanned aircraft systems (UAS). Beyond replacing human beings in aircraft that perform dangerous roles, UAS are highly valuable because they possess characteristics that many manned aircraft do not. For example, they can fly long-duration missions, thereby providing a sustained presence over the battlefield. Because of greater demand for UAS, the Department of Defense (DOD) continues to increase its investment in these programs, requesting approximately $6.1 billion in fiscal year 2010 for new systems and expanded capabilities in existing ones. In 2000, DOD had fewer than 50 unmanned aircraft in its inventory; as of October 2009, this number had grown to more than 6,800. Although each of the military services operates several types of UAS, the Air Force and the Army have acquired a greater number of larger, more capable systems that have been deployed to support ongoing operations.

While DOD has expanded its inventories of UAS to meet warfighter demand, our prior work has found that DOD has faced obstacles in overcoming challenges in the development and acquisition of UAS programs and in the integration of these systems into combat operations.[1] For example, in 2007 we reported that because DOD began the UAS acquisition process too early, the related UAS development plans

---

[1] See, for example, GAO, *Defense Acquisitions: Greater Synergies Possible for DOD's Intelligence, Surveillance, and Reconnaissance Systems*, GAO-07-578 (Washington, D.C.: May 17, 2007); *Unmanned Aircraft Systems: Advance Coordination and Increased Visibility Needed to Optimize Capabilities*, GAO-07-836 (Washington, D.C.: July 11, 2007); *Unmanned Aircraft Systems: Additional Actions Needed to Improve Management and Integration of DOD Efforts to Support Warfighter Needs*, GAO-09-175 (Washington, D.C.: Nov. 14, 2008); and *Defense Acquisitions: Opportunities Exist to Achieve Greater Commonality and Efficiencies among Unmanned Aircraft Systems*, GAO-09-520 (Washington, D.C.: July 30, 2009).

contained requirements and funding uncertainties. We also reported in 2007 that DOD had been unable to fully optimize the use of its UAS assets in combat operations because it lacked an approach to allocating and tasking them that considered the availability of all assets in determining how best to meet warfighter needs. In 2008, we reported that DOD had not developed a comprehensive and integrated strategic plan with priorities, timelines, and long-term implementation goals to align departmental and military service efforts in order to improve the management and operational use of UAS. More recently, the Congress has expressed interest in DOD's plans regarding UAS, for example, in the steps that DOD has taken to develop qualifications for UAS operators necessary for the routine access of unmanned aircraft to U.S. airspace to conduct training and operations.

Integral to the operation of UAS are numerous support elements—including personnel, facilities, and a communications infrastructure to relay signals to and from the aircraft; programs to train personnel for UAS operations; and publications to guide personnel as they conduct training and operations. Regarding training programs, DOD guidance directs the military services to take actions to support joint and integrated operations training to the maximum extent possible.[2] Thus, training programs ideally require access to the national airspace system (a complex system comprising thousands of people, procedures, facilities, and pieces of equipment) and opportunities for ground combat units and UAS personnel to participate in joint training exercises so that these personnel can practice the interactions they will have with one another on the battlefield. However, DOD's UAS operations are subject to numerous restrictions,[3] which can create competition for the limited available airspace and can constrain DOD's ability to effectively utilize training and operational locations. Further, commitments to ongoing operations can limit the amounts of UAS personnel and equipment that are available to conduct training. Because of airspace access and personnel and equipment availability issues, DOD has used simulators (or virtual training devices) to increase training opportunities. To guide service and joint training programs and to assist individuals and units in integrating military capabilities in joint operations, the military services are responsible for

---

[2] Department of Defense Directive 1322.18, *Military Training* (Jan. 13, 2009).

[3] UAS training operations are generally restricted to DOD-designated airspace because current systems do not meet several federal requirements. For example, UAS do not have personnel or a suitable alternative technology on board to detect and avoid other aircraft.

coordinating with each other to develop timely publications. These publications describe doctrine, tactics, techniques, procedures, and concepts of operations and can be used to optimize the integration of UAS during joint operations.

As you requested, we evaluated DOD's ability to support UAS inventories. Specifically, we determined the extent to which (1) plans were in place to account for the personnel, facilities, and communications infrastructure needed to support Air Force and Army UAS inventories; (2) DOD addressed challenges that affect the ability of the Air Force and the Army to train personnel for UAS operations; and (3) DOD updated its existing publications that articulate doctrine and tactics, techniques, and procedures to reflect the knowledge gained from using UAS in ongoing operations.

To determine the extent to which plans were in place to account for the personnel, facilities, and communications infrastructure to support Air Force and Army UAS inventories, we focused primarily on Air Force and Army UAS programs that support ongoing operations. Excluded from this review were programs for small unmanned aircraft. While the military services have acquired more than 6,200 of these aircraft, they generally do not have substantial support requirements. We examined UAS program and funding plans and DOD's policies governing the requirements definition and acquisition processes. We consulted the Office of Management and Budget's *Capital Programming Guide* and our *Cost Estimating and Assessment Guide* for instruction on developing cost estimates and plans to manage capital investments.[4] In determining the extent to which DOD addressed challenges that affect the ability of the Air Force and the Army to train personnel for UAS operations, we visited select military installations and the Army's National Training Center at Fort Irwin, California, and spoke with knowledgeable DOD officials to determine the specific challenges that the Air Force and the Army faced when training service personnel to perform UAS missions in joint operations. Specifically, we spoke with personnel in Air Force and Army UAS units in the United States and in Iraq to identify the training they were able to perform prior to operating UAS in joint operations and the

---

[4] See Office of Management and Budget, *Capital Programming Guide: Supplement to Circular A-11, Part 7, Planning, Budgeting, and Acquisition of Capital Assets* (Washington, D.C.: June 2006), and GAO, *GAO Cost Estimating and Assessment Guide: Best Practices for Developing and Managing Capital Program Costs*, GAO-09-3SP (Washington, D.C.: March 2009).

challenges, if any, that prevented them from performing their required training tasks. In identifying Air Force and Army unit personnel to speak with, we selected a nonprobability sample of units that were preparing to deploy for contingency operations or had redeployed from these operations from May 2009 through September 2009. We assessed DOD's efforts to overcome these challenges in light of leading practices derived from principles established under the Government Performance and Results Act of 1993 and key elements of an overarching organizational framework, such as developing results-oriented strategies, as described in our prior work.[5] To determine the extent to which DOD had updated its existing publications that articulate doctrine and tactics, techniques, and procedures to reflect the knowledge gained from using UAS in ongoing operations, we reviewed joint, multiservice, and service-specific UAS doctrine, tactics, techniques, procedures, and concepts of operations. We interviewed DOD and military service officials and analyzed publications to determine how the documents articulate knowledge gained from using UAS in ongoing operations; the degree to which information is provided for UAS stakeholders, such as military planners and ground commanders; and the processes that the services use to update the publications. We conducted this performance audit from October 2008 through March 2010 in accordance with generally accepted government auditing standards. Those standards require that we plan and perform the audit to obtain sufficient, appropriate evidence to provide a reasonable basis for our findings and conclusions based on our audit objectives. We believe that the evidence obtained provides a reasonable basis for our findings and conclusions based on our audit objectives. A more detailed discussion of our scope and methodology is provided in appendix I.

Background

DOD defines a UAS as a system whose components include the necessary equipment, networks, and personnel to control an unmanned aircraft—that is, an aircraft that does not carry a human operator and is capable of flight under remote control or autonomous programming. Battlefield commanders have experienced a high level of mission success in ongoing

---

[5] See, for example, GAO, *Highlights of a GAO Roundtable: The Chief Operating Officer Concept: A Potential Strategy to Address Federal Governance Challenges*, GAO-03-192SP (Washington, D.C.: Oct. 4, 2002); *Highlights of a GAO Forum: Mergers and Transformation: Lessons Learned for a Department of Homeland Security and Other Federal Agencies*, GAO-03-293SP (Washington, D.C.: Nov. 14, 2002); *Defense Business Transformation: Achieving Success Requires a Chief Management Officer to Provide Focus and Sustained Leadership*, GAO-07-1072 (Washington, D.C.: Sept. 5, 2007); and GAO-09-175.

operations with capabilities provided by UAS. Beyond a traditional intelligence, surveillance, and reconnaissance role, UAS have been outfitted with missiles to strike targets, with equipment to designate targets for manned aircraft by laser, and with sensors to locate the positions of improvised explosive devices and fleeing insurgents, among other tasks.

DOD has acquired UAS through formal acquisition programs, and in certain cases, the military services have purchased common UAS components. For example, the Army and the Marine Corps are purchasing the Shadow UAS and the Air Force and the Navy are acquiring a similar unmanned aircraft for the Global Hawk and the Broad Area Maritime Surveillance UAS programs. DOD has also fielded other UAS in order to meet urgent warfighter requests and for technology demonstrations. In 2008, U.S. Joint Forces Command's Joint UAS Center of Excellence established a system to categorize UAS in groups that are based on attributes of vehicle airspeed, weight, and operating altitude. For example, group 1 UAS weigh 20 pounds or less whereas group 5 UAS weigh more than 1,320 pounds. Table 1 provides the military services' inventories of groups 3, 4, and 5 unmanned aircraft as of October 2009.

Table 1: Military Services' Inventories of Selected Unmanned Aircraft

Military service	Group	System	Number of aircraft
Air Force	4	Predator	140
	5	Global Hawk	17
	5	Reaper	35
	Total		**192**
Army	3	Shadow	288
	4	Extended Range Multi-Purpose	4
	4	Fire Scout	32
	4	Hunter	22
	4	Warrior	18
	Total		**364**
Navy	4	Fire Scout	7
	5	Global Hawk Maritime Demonstration	2
	5	Reaper	4
	5	Unmanned Combat Air System	2
	Total		**15**
Marine Corps	3	Shadow	28
	Total		**28**

Source: GAO analysis of DOD data.

Note: The military services have also acquired more than 6,100 group 1 unmanned aircraft, such as the Raven, and more than 100 group 2 unmanned aircraft, such as the Scan Eagle. These systems were excluded from this review because smaller UAS generally do not have substantial support requirements.

Several major systems—including the Air Force Predator, Reaper, and Global Hawk; the Army and Marine Corps Shadow; and the Army Extended Range Multi-Purpose (ERMP) UAS—have been deployed and used successfully in combat. Because of the resulting demand for these assets, several of the military services' UAS programs have experienced significant growth. For example, DOD's fiscal year 2010 budget request sought funds to continue to increase the Air Force's Predator and Reaper UAS programs to 50 combat air patrols by fiscal year 2011—an increase of nearly 300 percent since fiscal year 2007.[6] DOD's fiscal year 2007 through

---

[6] DOD's February 2010 *Quadrennial Defense Review Report* states that the Air Force is on track to achieve this goal and that it will continue to increase the number of combat air patrols to 65 by fiscal year 2015.

fiscal year 2010 budget requests for all of DOD's UAS programs reflect an increase in the amount of funding requested by DOD for UAS investments to support warfighting needs, as shown in table 2.

Table 2: DOD's Budget Requests for UAS (Fiscal Years 2007 through 2010)

In fiscal year 2009 constant dollars in millions

	2007	2008	2009	2010	Total
Research, development, test and evaluation	$1,778.9	$1,668.3	$2,016.4	$2,519.6	**$7,983.1**
Procurement	2,201.4	2,968.3	3,372.2	3,596.8	**$12,138.7**
Total	**$3,980.3**	**$4,636.6**	**$5,388.6**	**$6,116.4**	**$20,121.8**

Source: GAO analysis of funding requests for UAS included in the President's fiscal year 2009 and fiscal year 2010 budget requests, including funds to support contingency operations.

Note: Numbers may not add to totals due to rounding.

Beyond development and acquisition costs, DOD's UAS programs have additional funding requirements, for example, those costs to operate and sustain the weapon system, to provide personnel, and to construct facilities and other infrastructure. DOD guidance encourages acquisition personnel to consider factors, including personnel, facilities, supporting infrastructure, and policy costs, when fielding new capabilities.[7] However, DOD's and our prior work have found that decision makers have had limited visibility over total weapon system costs because estimates have not reflected a full accounting of life cycle costs. In a November 2009 report, for example, DOD concluded that its acquisition processes pay too little attention to weapon system support costs, even though the department spends more than $132 billion each year to sustain its weapon systems.[8] The report also concluded that the lack of adequate visibility of operating and support costs has been a long-standing barrier to effectively assessing, managing, and validating the benefits or shortcomings of support strategies. In our prior work, we have found that DOD often makes inaccurate funding commitments to weapon system programs based on unrealistic cost estimates.[9] The foundation of an accurate

---

[7] Chairman of the Joint Chiefs of Staff, *Manual for the Operation of the Joint Capabilities Integration and Development System* (July 31, 2009), cited in Chairman of the Joint Chiefs of Staff Instruction 3170.01G, *Joint Capabilities Integration and Development System* (Mar. 1, 2009), https://acc.dau.mi/pm (accessed Feb. 1, 2010).

[8] Department of Defense, *Weapon System Acquisition Reform Product Support Assessment* (November 2009).

[9] GAO, *Defense Acquisitions: A Knowledge-Based Funding Approach Could Improve Major Weapon System Program Outcomes*, GAO-08-619 (Washington, D.C.: July 2, 2008).

funding commitment should be a realistic cost estimate that allows decision makers to compare the relative value of one program to another and to make adjustments accordingly. We reported that DOD's unrealistic cost estimates were largely the result of a lack of knowledge, failure to adequately account for risk and uncertainty, and overly optimistic assumptions about the time and resources needed to develop weapon systems. By repeatedly relying on unrealistically low cost estimates, DOD has initiated more weapon systems programs than its budget can afford.

We have also conducted an extensive body of work on DOD's efforts to ensure the availability of defense critical infrastructure, which includes space, intelligence, and global communications assets, reporting on DOD's progress in addressing the evolving management framework for the Defense Critical Infrastructure Program, coordination among program stakeholders, implementation of key program elements, the availability of public works infrastructure, and reliability issues in DOD's lists of critical assets, among other issues.[10] For example, we reported in 2008 on the challenges that the Air Force faced in addressing the continuity of operations and physical security at Creech Air Force Base, a location where nearly half of the Air Force's UAS operations were being performed at the time.[11]

While many of DOD's UAS operations currently take place outside of the United States, primarily in Iraq and Afghanistan, the military services require access to the national airspace system to conduct UAS training, among other reasons, and personnel and equipment to support training exercises. However, DOD has experienced several challenges in gaining access to the national airspace system and limitations in the availability of UAS personnel and equipment to support training because of operational commitments. Because DOD's UAS do not meet several federally mandated requirements for routine access to the national airspace system, most types of UAS may not perform routine flight activities, such as taking

---

[10] See, for example, GAO, *Defense Critical Infrastructure: DOD's Evolving Assurance Program Has Made Progress but Leaves Critical Space, Intelligence, and Global Communications Assets at Risk*, GAO-08-828NI (Washington, D.C.: Aug. 22, 2008), and *Defense Critical Infrastructure: Actions Needed to Improve the Identification and Management of Electrical Power Risks and Vulnerabilities to DOD Critical Assets*, GAO-10-147 (Washington, D.C.: Oct. 23, 2009).

[11] GAO, *Defense Critical Infrastructure: Additional Air Force Actions Needed at Creech Air Force Base to Ensure Protection and Continuity of UAS Operations*, GAO-08-469RNI (Washington, D.C.: Apr. 23, 2008).

off and landing outside DOD-managed airspace. For example, UAS do not have personnel or a suitable alternative technology on board the aircraft to detect, sense, and avoid collision with other aircraft. The Federal Aviation Administration approves applications from DOD (and other government agencies) for authority to operate UAS in the national airspace system outside of that restricted for DOD's use on a case-by-case basis.

To provide military personnel with information on UAS, DOD components, which include the military services and other defense organizations, have produced several publications, including joint and service doctrinal publications that describe processes to plan for and integrate UAS into combat operations. In addition, DOD components have produced concepts of operations for UAS, as well as multiservice and platform-specific tactics, techniques, and procedures manuals. These publications are intended to provide planners at operational and tactical levels of command, such as joint task forces and divisions, with an understanding of the processes to incorporate UAS into their intelligence collection plans and into combat operations. Tactical ground units requesting support from UAS, which can range from small special operations units to large infantry brigades engaged in ground combat operations, may use these documents to understand UAS capabilities and how to best incorporate them into preplanned and dynamic missions. UAS operators use these documents to establish best practices, standard operating procedures for integrating UAS into joint operations, and processes for interacting with other air and ground forces on the battlefield. Periodically, DOD components update these publications to include new knowledge on military practices and capabilities. Generally, these updates are accomplished through comprehensive service- or departmentwide reviews conducted by subject matter experts.

Plans Are Not in Place to Fully Account for the Personnel, Facilities, and Some Communications Infrastructure Needed to Support Air Force and Army UAS Programs

DOD has policies that encourage its components to plan for factors, including personnel, facilities, and communications infrastructure, that are needed to support weapon systems programs. Extensive planning for these factors provides decision makers with complete information on total program costs and assurances that weapon system programs can be fully supported in the long term. During our review, however, we identified areas where, despite the growth in UAS inventories, comprehensive plans for personnel, facilities, and some communications infrastructure have not been fully developed to support Air Force and Army UAS programs.

DOD Has Processes to Plan for Personnel, Facilities, and Communications Infrastructure for UAS Programs

DOD guidance recommends that acquisition personnel determine a weapon system program's life cycle costs by conducting planning for the manpower, facilities, and other supporting infrastructure, among other factors, needed to support a weapon system, and fully fund the program and manpower needed in budget requests.[12] Decision makers use this information to determine whether a new program is affordable and the program's projected funding and manpower requirements are achievable. DOD components are expected to conduct continuing reviews of their strategies to sustain weapon systems programs and to identify deficiencies in these strategies, making necessary adjustments to them in order to meet performance requirements.

In addition, the Office of Management and Budget's *Capital Programming Guide* also indicates that part of conducting cost analyses for capital assets, such as weapon systems, is refining cost estimates as programs mature and as requirements change, and incorporating risk analyses in these estimates.[13] We have reported that accurate cost estimates are necessary for government acquisition programs for many reasons, for example, to evaluate resource requirements, to support decisions about funding one program over another, and to develop annual budget

---

[12] See Department of Defense Instruction 5000.02, *Operation of the Defense Acquisition System* (Dec. 8, 2008), and Department of Defense, *Defense Acquisition Guidebook* (Washington, D.C.: Dec. 17, 2009), https://dag.dau.mil (accessed Jan. 5, 2010).

[13] Office of Management and Budget, *Capital Programming Guide*.

requests.[14] Moreover, having a realistic estimate of projected costs makes for effective resource allocations, and it increases the probability of a program's success.

Service Strategies Are Not Fully Developed to Supply the Personnel Needed to Support UAS Programs

The Air Force and the Army train personnel to perform functions for UAS operations, such as operating the aircraft and performing maintenance. Because of the rapid growth of UAS programs, the number of personnel required to perform these functions has substantially increased and the services have taken steps to train additional personnel. However, in service-level UAS vision statements, the Air Force and the Army have identified limitations in their approaches to provide personnel for UAS operations, but they have not yet fully developed strategies that specify the actions and resources required to supply the personnel needed to meet current and projected future UAS force levels.

The Air Force, for example, has identified limitations in the approaches it has used to supply pilots to support the expanded Predator and Reaper UAS programs. Since the beginning of these programs, the Air Force has temporarily reassigned experienced pilots to operate UAS, and more recently, it began assigning pilots to operate UAS immediately after they completed undergraduate pilot training. Air Force officials stated that this initiative is intended to provide an additional 100 pilots per year on a temporary basis to support the expanding UAS programs. While the Air Force has relied on these approaches to meet the near-term increase in demand for UAS pilots, officials told us that it would be difficult to continue these practices in the long term without affecting the readiness of other Air Force weapon systems, since the pilots who are performing UAS operations on temporary assignments are also needed to operate other manned aircraft and perform other duties.

In an attempt to develop a long-term, sustainable career path for UAS pilots, the Air Force implemented a new initiative in 2009 to test the feasibility of establishing a unique training pipeline for UAS pilots. Students selected for this pipeline are chosen from the broader Air Force officer corps and are not graduates of pilot training. At the time of our work, the Air Force was analyzing the operational effectiveness of those personnel who graduated from the initial class of the test training pipeline to determine if this approach could meet the long-term needs of the Air

---

[14] GAO-09-3SP.

Force. In addition, officials told us that the Air Force would ultimately need to make some changes to this pipeline to capture lessons learned from the initial training classes and to help ensure that graduates were effectively fulfilling UAS mission requirements. For example, officials stated that the initial graduates of the training pipeline have not yet been provided with training on how to take off and land the Predator and that these functions are being performed by more experienced pilots. However, the Air Force had neither fully determined the total training these personnel would require to effectively operate the Predator and Reaper aircraft during UAS missions nor fully determined the costs that would be incurred to provide training for these assignments. Officials estimated that it would take at least 6 months after the second class of personnel graduated from the training pipeline to assess their effectiveness during combat missions and to determine what, if any, additional training these personnel require.

Further, the Air Force has not finalized an approach to supply the personnel needed to perform maintenance functions on the growing UAS inventories and meet servicewide goals to replace contractor maintenance positions with funded military ones. Currently, the Air Force relies on contractors to perform a considerable portion of UAS maintenance because the Air Force does not have military personnel trained and available to perform this function. For example, contractors perform approximately 75 percent of organization-level maintenance requirements for the Air Combat Command's Predator and Reaper UAS. According to the Air Force's UAS *Flight Plan*,[15] replacing contractor maintenance personnel with military personnel would enable the Air Force to develop a robust training pipeline and to build a sustainable career field for UAS maintenance, while potentially reducing maintenance costs. According to officials with whom we spoke, the Air Force's goal is to establish a training pipeline for military maintenance personnel by fiscal year 2012. However, the Air Force has not developed a servicewide plan that identifies the number of personnel to be trained, the specific training required, and the resources necessary to establish a dedicated UAS training pipeline. Officials estimated that it could take until fiscal year 2011 to determine these requirements and to test the feasibility of a new training pipeline.

---

[15] Department of Defense, *United States Air Force Unmanned Aircraft Systems Flight Plan 2009-2047* (May 2009).

Our review also found that the Army's personnel authorizations are insufficient to fully support UAS operations. For example, according to officials, the Army has determined on at least three separate occasions since 2006 that Shadow UAS platoons did not have adequate personnel to support the near-term and projected pace of operations. Officials from seven Army Shadow platoons in the United States and in Iraq with whom we spoke told us that approved personnel levels for these platoons did not provide an adequate number of vehicle operators and maintenance soldiers to support continuous UAS operations. Army officials told us that currently approved personnel levels for the Shadow platoons were based on planning factors that assumed that the Shadow would operate 12 hours per day with the ability to extend operations to up to 16 hours for a limited period of time. However, personnel with these platoons told us that UAS in Iraq routinely operated 24 hours per day for extended periods of time. Army officials also told us that organizations, such as combat brigades and divisions, require additional personnel to provide UAS expertise to assist commanders in optimizing the integration of UAS into operations and safely employing these assets.

Despite the shortfalls experienced during ongoing operations, the Army has yet to formally increase personnel authorizations to support UAS operations or approve a servicewide plan to provide additional personnel. Officials told us that on the basis of these and other operational experiences, the Army was in the process of developing initiatives to provide additional personnel to Army organizations to address personnel shortfalls, and included these initiatives in an October 2009 UAS vision statement developed by the Army's UAS Center of Excellence. These initiatives include increasing authorized personnel levels for vehicle operators and maintenance soldiers in Shadow UAS platoons as well as other initiatives to assign UAS warrant officers and Shadow vehicle operators to brigade and division staffs. According to the Army's UAS vision statement, the initiatives to increase UAS personnel to meet current and projected requirements will be completed by 2014. However, at the time of our work, the Army had not developed a detailed action plan that identified the number of additional personnel that would support UAS operations and the steps it planned to take in order to synchronize the funding and manpower necessary to provide these personnel, such as reallocating existing manpower positions within combat brigades to increase the size of Shadow platoons.

Facilities Needed to Support UAS Programs Have Not Been Systematically Defined and Costs Are Uncertain

Although DOD has requested funding to some extent in recent budget requests and expects to request additional funds in future years, the Air Force and the Army have not fully determined the specific number and type of facilities needed to support UAS training and operations. For example:

- The Air Force has neither determined the total number of facilities required to support its rapidly expanding Predator and Reaper programs nor finalized the criteria it will use to renovate existing facilities because decisions regarding the size of UAS squadrons and the locations where these squadrons will be based had not been finalized. In some cases, the Air Force has constructed new facilities to support UAS operations. In other cases, the Air Force determined that it did not need to construct new facilities and is instead renovating existing facilities on UAS operating locations, such as maintenance hangars and buildings to use for unit operations facilities. However, until the Air Force determines where it plans to locate all of its new UAS units and finalizes the criteria that would be used to guide the construction or renovation of facilities, the Air Force will be unable to develop realistic estimates of total UAS facility costs and long-term plans for their construction.
- The Army has begun to field the ERMP UAS and has determined that the Army installations where the system will be stationed require facilities uniquely configured to support training and operations. These facilities include a runway, a maintenance hangar, and a unit operations facility. However, the Army has not fully determined where it will base each of these systems and it has not completed assessments at each location to evaluate existing facilities that could potentially be used to meet the ERMP requirements and to determine the number of new facilities that the Army needs to construct. The lack of detailed facility planning has affected the Army's fielding schedule for the ERMP. Army officials told us that the fielding plan for this system has been adjusted to give priority to locations that do not require significant construction. According to Army officials, initially the Army had developed its fielding plan for the ERMP so that the plan for fielding the system synchronized with the estimated deployment dates for units supporting ongoing contingency operations.
- The Army has not definitively determined, for the Shadow UAS, the type and number of facilities needed to support training and aircraft storage. In 2008, the Army established a policy that directed its ground units to store Shadow aircraft in facilities with other ground unit tactical equipment and not in facilities uniquely configured for these

aircraft.[16] Ground units typically store equipment in facilities, such as motor pools, that are not always near training ranges. Previously, the Army had allowed some units to construct unique facilities for the Shadow nearby installation ranges to facilitate their ability to conduct training. Army officials told us that storing equipment within the motor pool creates constraints to training when ranges are not in proximity. In these situations, units are required to transport the Shadow and its associated equipment from the motor pool to the training range, assemble and disassemble the aircraft, and transport the equipment back to the motor pool. Officials we spoke with at one Shadow platoon estimated that these steps required more than 3 hours to complete, thereby limiting the amount of flight training that can be performed during one day. This practice may also lead to a more rapid degradation of aircraft components. Officials told us that the frequent assembling and disassembling of aircraft increases the wear and tear on components, which could increase maintenance costs. While the Army maintains a process for installations to request a waiver from the policy that would allow for the construction of unique aircraft facilities, officials told us that the Army is reevaluating whether the Shadow requires unique facilities. Any decision to change the policy on Shadow facilities would ultimately increase total program costs.

Because systematic analyses of facility needs for UAS programs have not been conducted, the total costs to provide facilities for Air Force and Army UAS programs are uncertain and have not been fully accounted for in program cost estimates that are used by decision makers to evaluate the affordability of these programs. Further, although costs for facilities were not included in these estimates, our analysis of DOD's budget requests for fiscal year 2007 through fiscal year 2010 found that the Air Force and the Army have sought more than $300 million to construct facilities for UAS. Moreover, as these services finalize assessments of the number and type of facilities required for UAS operations and field additional systems, they will likely request additional funds for facilities. For example, Army officials told us that cost estimates for ERMP facilities would be unavailable until all of the ongoing requirements assessments were complete; however, our analysis of the Army's facility plans for the ERMP

---

[16] In contrast, the Marine Corps, which also operates the Shadow UAS, has determined that the system has a facility requirement. The Marine Corps has requested military construction funds to build new facilities to support its systems.

estimates that the Army could request more than $600 million to construct facilities for this program alone.[17]

The Air Force Does Not Have a Plan in Place to Address Near-Term Risks to Communications Infrastructure

In general, the military services operate UAS using two different operational concepts. For example, Army and Marine Corps units primarily conduct UAS operations through a line-of-sight operational concept. As depicted in figure 1, UAS are launched, operated, and landed in this concept nearby the ground units that they support and are controlled by a ground station that is also nearby.

---

[17] This estimate is based on our analysis of the notional facility requirement for an ERMP UAS to include a maintenance hangar, a company operations facility, and a landing surface for fielding the system to 10 combat aviation brigades.

Figure 1: Line-of-Sight UAS Operational Concept

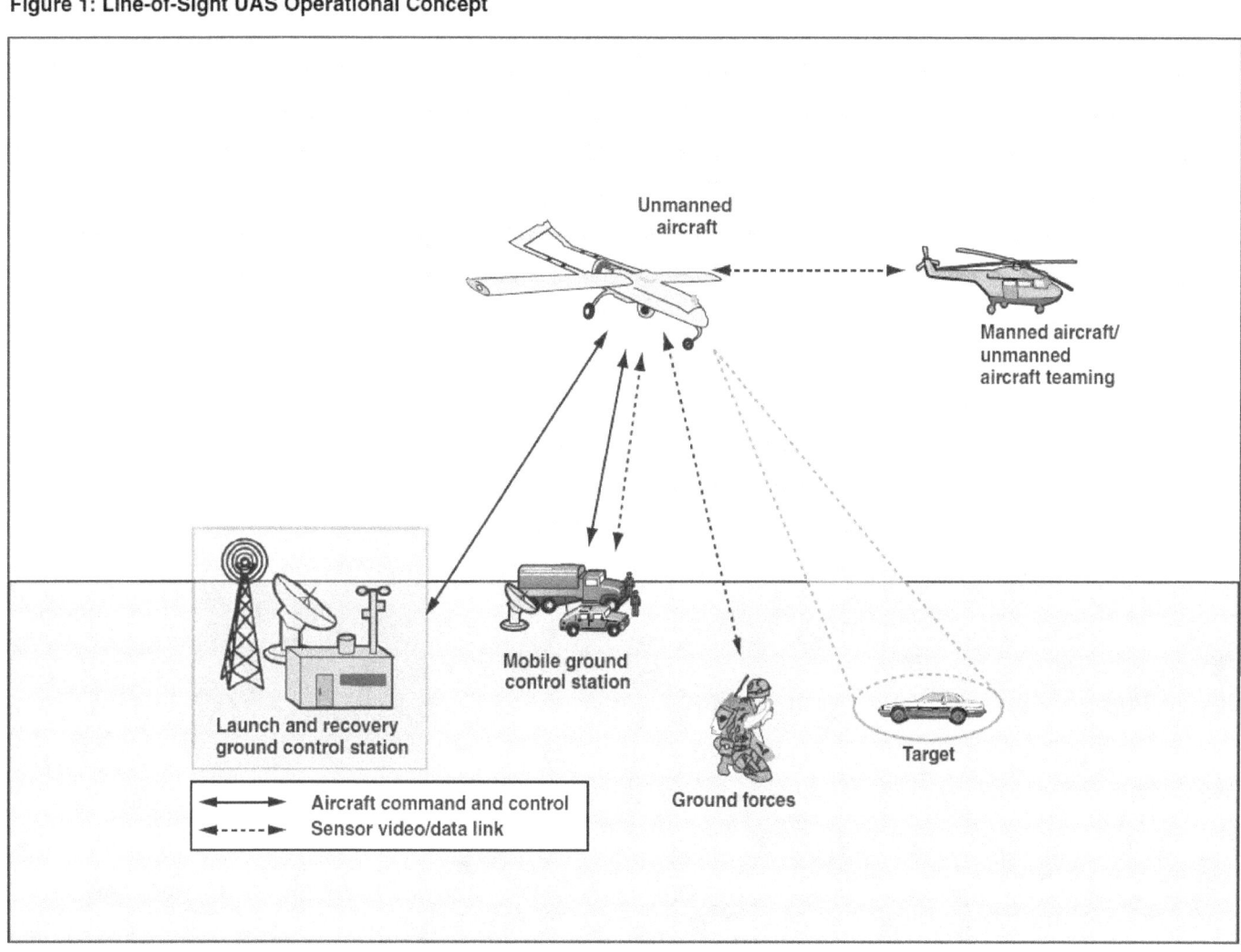

Sources: GAO analysis of DOD data; Art Explosion (Images).

In this concept, UAS can also transmit video and data to ground units or other aircraft within line of sight to support a range of missions, such as reconnaissance, surveillance, and target acquisition. Some level of risk is introduced in a line-of-sight operational concept if the command and control links to the aircraft are not secure.

Air Force and Navy units use this line-of-sight concept but also use a beyond-the-line-of-sight operational concept that increases the risk of a disruption in operations. In this concept, the operation of the UAS relies on additional equipment and networks, some of which are located outside of the country where the UAS operations occur. According to Air Force officials, the use of a beyond-the-line-of-sight concept permits the service to conduct UAS operations with limited numbers of personnel and equipment deployed within an operational theater. As in the line-of-sight concept, the UAS are launched and landed by deployed ground control stations; however, the UAS are controlled during missions by a pilot and sensor operator located at a fixed ground control station located at a remote site. A satellite relay site delivers the signals between the UAS and the ground control station at the remote site (see fig. 2).

Figure 2: Beyond-the-Line-of-Sight UAS Operational Concept

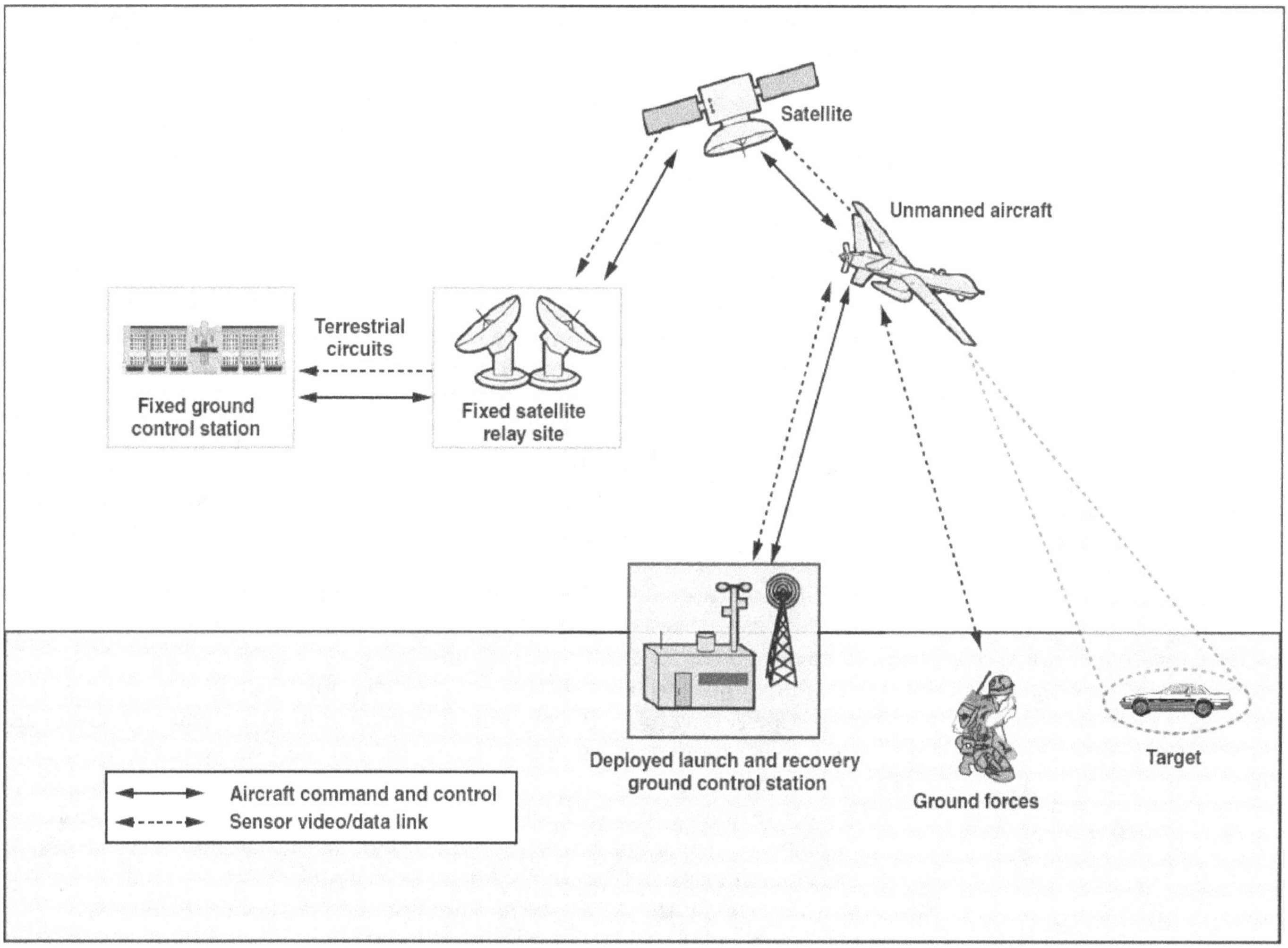

Sources: GAO analysis of DOD data; Art Explosion (Images).

The Air Force currently employs this operational concept for Predator, Reaper, and Global Hawk UAS missions that support contingency operations in Iraq and Afghanistan. For these missions, a ground control station located within the United States takes control of the aircraft. A satellite relay site at a fixed location (located outside of the continental United States) relays signals from the ground control station to the UAS so that they can communicate.[18] Any disruptions at the satellite relay site caused, for example, by a natural or man-made disaster could affect the number of UAS operated under this concept.

DOD assesses risks and vulnerabilities to its critical assets and installations using the Defense Critical Infrastructure Program and other mission assurance programs and efforts, including those related to force protection, antiterrorism, continuity of operations, and installation preparedness.[19] For example, Air Force doctrine dated June 2007 calls for the establishment of backup or redundant command and control systems for high-value systems so that operations can continue in the event of failure or damage of the primary system.[20] This doctrine further states that planning for redundant command and control systems should be formalized and exercised before military operations begin. However, the Air Force has not established an alternate, redundant satellite relay site with the capacity to control all UAS missions that are supporting ongoing combat operations. Because of the satellite relay's critical importance in supporting ongoing contingency operations, the Air Force is taking steps to establish a redundant satellite relay site to support UAS missions in the event of disruptions at the current location. For example, officials told us that the Air Force is acquiring new communications equipment with increased capacity for the current site, which will allow equipment currently in use to be available for other locations. In addition, the Air Force is seeking funds to conduct surveys to identify potential locations to establish a redundant satellite relay site. However, officials stated that these efforts are not scheduled to be completed until fiscal year 2012, at the earliest. Air Force officials also told us that they would have options to pursue in the event of a near-term disruption at the satellite relay site, such as relocating assets from other Air Force operations. At the time of our work, however, the Air Force had not conducted a detailed analysis of these options to determine the extent to which they would provide for the

---

[18] In addition, the Navy's Global Hawk Maritime Demonstration unmanned aircraft are controlled through the same location.

[19] As discussed earlier in this report, our prior work has identified a number of challenges that DOD faces with the evolving management framework of the Defense Critical Infrastructure Program. See, for example, GAO-08-828NI and GAO-10-147.

[20] Department of Defense, *Air Force Doctrine Document: Command and Control 2-8* (June 2007).

continuity of UAS operations, or established a specific milestone to formalize a plan that could be implemented quickly in the event of a disruption.

Various Factors Have Contributed to a Lag in Planning for Personnel, Facilities, and Communications Infrastructure for UAS Programs

Several factors have contributed to a lag in Air Force and Army planning for the personnel, facilities, and some communications infrastructure that are integral to the operation of UAS. For example, although DOD's primary requirements definition process—termed the Joint Capabilities Integration and Development System—encourages acquisition personnel to develop cost estimates for its new weapon systems programs, including consideration of various support factors, the Air Force's current UAS programs were, for the most part, initially developed and fielded as technology demonstrations. According to the Air Force, these programs have been subsequently approved within the Joint Capabilities Integration and Development System, but comprehensive life cycle plans that fully account for the personnel, facilities, and communications infrastructure to effectively manage the systems have not yet been completed.

Further, to meet near-term warfighter demands for these capabilities, several UAS programs have been expanded beyond planned force structure levels and, in some cases, have been fielded more rapidly than originally planned. Given the changes in program requirements in the near term, the Air Force and the Army have, for example, in the case of the Air Force Predator and the Army Shadow programs, taken measures to support UAS inventories. However, these measures have been taken without the benefit of rigorous planning for the specific numbers and types of personnel and facilities and some communications infrastructure that are needed to support these programs in the long term. Finally, while DOD components are expected to identify deficiencies in their strategies to support weapon systems programs and to make necessary adjustments to them as requirements change, the Air Force and the Army have not completed the analyses or developed plans to account for new personnel and facility requirements, and the Air Force has not developed a plan to ensure the communications infrastructure needed to support its UAS programs. In the absence of detailed action plans that fully account for these factors and include milestones for tracking progress and synchronize funding and personnel, DOD cannot have a reasonable assurance that these services' approaches will fully support current and projected increases in UAS inventories. In addition, the lack of comprehensive plans limits the visibility of decision makers to evaluate the total resources required to support UAS inventories and to make informed choices about funding one program over another.

DOD Has Not Resolved Challenges That Affect the Ability of the Air Force and the Army to Train Personnel for UAS Operations

Prior work shows that in order to improve the management of federal activities, it is important that agencies develop comprehensive strategies to address challenges that threaten their ability to meet long-term goals. We identified several initiatives that DOD has commenced to address UAS training challenges, but DOD lacks a results-oriented strategy to ensure that compatible goals and outcomes are achieved among these initiatives.

Availability of Airspace Limits Training Opportunities

Many of DOD's UAS operations take place outside of U.S. airspace, but DOD requires access to the national airspace system for training, to conduct operations such as homeland defense, and for the transit of unmanned aircraft to overseas deployment locations—requirements that have created airspace access challenges. For example, according to Army officials, a single Shadow UAS platoon requires more than 3,000 flight hours per year to fully train all aircraft operators. Because UAS do not meet various federally mandated requirements and therefore do not have routine access to the national airspace system, personnel must train in DOD-managed airspace and training ranges located near their home stations. Competing for this finite airspace are other units located at home stations that also require access to DOD-managed airspace for their operations, such as manned aircraft training. This competition, among other factors, has affected the amount of training UAS personnel can conduct and their ability to prepare for deployments. Army officials with four of the seven Shadow platoons we met with told us that they were unable to fully train the number of personnel needed to perform continuous combat missions before they deployed for overseas operations. As a result, UAS personnel had to conduct additional training tasks upon arrival in Iraq and Afghanistan.

Plans to further increase UAS inventories on selected military installations will likely further increase the demand for airspace. For example, the Army plans to increase the number of Shadow UAS from about 70 systems fielded at the time of our review to a goal of more than 100 systems by fiscal year 2015. According to current plans, all active and reserve component combat brigades, Army Special Forces units, fires brigades, and battlefield surveillance brigades will be provided with Shadow systems. In some cases, relocations of UAS to different installations have resulted in increased UAS inventories at the new installations. For example, in 2009, the Army moved the 4th Infantry Division and two combat brigades from Fort Hood, Texas, to Fort Carson, Colorado. This move resulted in the addition of two Shadow systems on Fort Carson. Army officials acknowledged that increases in UAS inventories will further complicate the competition for limited quantities of DOD-managed airspace.

As more advanced UAS are fielded in greater numbers, the military services will require increased access to the national airspace system. For example, the Army has fielded the ERMP UAS to its training battalion at Fort Huachuca, Arizona, and plans to provide one system, comprising 12 aircraft, to each of its active component combat aviation brigades. Because these aircraft are designed to operate at higher altitudes and possess capabilities beyond those on the Shadow UAS, officials told us that personnel who are responsible for operating the ERMP will require access to airspace that they cannot currently access to conduct training. Similarly, the Air Force requires expanded access to the national airspace system to train pilots who operate its UAS, and also to move aircraft, such as the Global Hawk, from bases in the United States to operational theaters around the world. Because UAS do not possess "sense and avoid" technology mandated by federal requirements for safe and efficient operations, the military services must provide, in many cases, an air- or ground-based observer of the aircraft during its flight in the national airspace system. According to DOD and military service officials, this restriction negates many of the most effective advantages of UAS, such as aircraft endurance, and creates an impractical requirement given the numbers of aircraft and personnel that are needed to monitor the unmanned aircraft during training. Moreover, the practice may be an unsustainable solution for meeting the demands of the military services' growing inventories of UAS. DOD estimated in a December 2008 report that based on planned UAS inventories in fiscal year 2013, the services will require more than 1 million flight hours to train UAS personnel within the United States.[21]

In recent years, DOD has taken several actions to integrate UAS into the national airspace system. For example, in November 2004, DOD issued an

---

[21] According to a DOD official, in February 2010 U.S. Joint Forces Command plans to publish revised estimates of annual flight hours required for UAS training. DOD's preliminary analysis of these estimates indicates a decrease in the number of flight hours needed to accomplish annual UAS training requirements.

airspace integration plan for unmanned aviation.[22] The plan established timelines and program milestones to achieve a goal that DOD's UAS would have safe, routine use of the national airspace system by 2010 while maintaining an equivalent level of safety to that of an aircraft with a pilot on board. In 2007, DOD convened a UAS Task Force with the participation of the Federal Aviation Administration and the Department of Homeland Security to find solutions to overcome the restrictions that limit the integration of UAS in the national airspace system, among other tasks. According to an official with the task force, DOD is in the process of revising the airspace integration plan by October 2010 to include near-, mid-, and long-term actions that DOD can take in concert with other federal agencies to improve the integration of UAS in the national airspace system. In our prior work, however, we reported that although some progress has been made to provide increased access to the national airspace system for small UAS, routine access for all types of UAS may not occur for a decade or more.[23]

The Congress has also raised questions about the progress made by DOD and other federal agencies in developing an approach to enable greater access for the department's UAS to the national airspace system. In the National Defense Authorization Act for Fiscal Year 2010, the Congress directed DOD and the Department of Transportation to jointly develop a plan to provide the military services' UAS with expanded national airspace system access. The plan, which is due April 2010, is to include recommendations concerning policies for the use of the national airspace system and operating procedures that should be implemented by both DOD and the Department of Transportation to accommodate UAS assigned to any state or territory of the United States.[24]

---

[22] Department of Defense, *Airspace Integration Plan for Unmanned Aviation* (November 2004).

[23] GAO, *Unmanned Aircraft Systems: Federal Actions Needed to Ensure Safety and Expand Their Potential Uses within the National Airspace System*, GAO-08-511 (Washington, D.C.: May 15, 2008).

[24] Pub. L. No. 111-84, § 935 (2009).

Limited Opportunities Exist for Air Force and Army Units to Train Together in a Joint Environment and Available Training Opportunities Have Not Maximized the Use of UAS

Army ground combat units and Air Force UAS units primarily train together at the Army's large training centers and not at home stations. In the United States, the Army has two large training centers—the National Training Center at Fort Irwin, California, and the Joint Readiness Training Center at Fort Polk, Louisiana. Army ground combat units conduct 2-week mission rehearsal exercises at one of these training centers before deploying for ongoing operations. The Air Force, however, has UAS stationed in the United States only near the National Training Center, so Air Force UAS do not support Army training exercises at the Joint Readiness Training Center.[25]

At the National Training Center, several factors limit the time Air Force UAS are available to support ground unit training. First, considerable numbers of Air Force UAS personnel and equipment items are supporting overseas contingency operations and therefore are unavailable to participate in training exercises in a joint environment. Air Force officials with the 432nd Wing, the unit that operates Air Force's Predator and Reaper UAS, told us that all of its unmanned aircraft are deployed to support overseas operations except for those that are supporting the initial training of UAS personnel or the testing of aircraft. These officials stated that in the event that additional aircraft were made available, the wing's personnel levels are insufficient to support additional training events because the unit does not have adequate personnel to support projected operational commitments and greater numbers of training exercises. Second, Army and Air Force officials told us that when Air Force UAS are at the training center, these aircraft are not always available to support ground unit training because a considerable portion of the UAS flight time is dedicated to accomplishing Air Force crewmember training tasks. Officials told us that the Army and Air Force have reached an informal agreement to allot about half of the time that an Air Force UAS is flying at the training center to support Army ground unit training objectives and the other half to accomplish Air Force training tasks. Air Force officials pointed out that although they try to align their crewmember training syllabi with ground unit training objectives at the National Training Center, training new personnel to operate these aircraft is their priority. Third, UAS may not be available during certain hours to support ground

---

[25] Officials pointed out that because of the beyond-the-line-of-sight operational concept, Air Force UAS stationed at current bases are capable of supporting training at the Joint Readiness Training Center; however, challenges associated with gaining access to the airspace needed to transit to Fort Polk make it impractical to participate in exercises at the training center.

unit training, which can occur on a 24-hour schedule. For example, Predator UAS from the California Air National Guard are available to support ground units only during daylight hours. To travel to the training center, these aircraft must pass through segments of national airspace that are not restricted for DOD's use and therefore must rely on a ground-based observer or on chase aircraft to follow them to and from the training center. Because of this reliance on ground or airborne observers, flights to and from the training center must be accomplished during daylight hours.

As a result of the limited number of unmanned assets that are available to support ground unit training at the National Training Center and the Joint Readiness Training Center, Army ground units conducting training exercises have relied on manned aircraft to replicate the capabilities of the Air Force's Predator and Reaper UAS. Officials told us that the use of manned aircraft in this role permits ground units to practice the process to request and integrate the capabilities provided by Air Force UAS in joint operations. However, this practice is not optimal as the manned aircraft do not replicate all of the capabilities of the Predator and Reaper aircraft, such as longer dwell times. At the time of our work, DOD was analyzing the utilization of manned aircraft for this purpose in order to assess whether there is a need for additional UAS to support joint training.

Additionally, when UAS are available to support ground unit training, we found that several factors affect the ability of ground combat units to maximize the use of available assets during training exercises. Officials we spoke with at the National Training Center pointed out that the effective integration of UAS in training exercises, like the integration of other types of joint air assets, depends on the priority that ground units place on developing training objectives that require the participation of joint air assets and their ability to plan for the use of these assets in the exercise. An Army Forces Command official stated that Army combat brigades often focus UAS training objectives during exercises on integrating their Shadow UAS and do not emphasize planning for and employing Air Force UAS. This is consistent with challenges that DOD has found in the integration of other joint air assets with ground unit training at the Army's training centers. A 2009 U.S. Joint Forces Command study found that although the National Training Center provides well-designed training environments to integrate Air Force aviation assets to support combat brigade training, a lack of adequate pre-exercise planning resulted in aircraft that were not fully integrated with ground combat units in training scenarios.[26] The study recommended that to improve the integration of joint air assets into ground training, ground units should conduct planning meetings with Air Force organizations early in the training process to identify mutually supporting training objectives and to synchronize air assets to achieve these training objectives.

Air Force and Army UAS Simulators Have Limited Capabilities to Enhance Training, and Long-Term Plans Are Unclear

DOD officials have indicated that UAS simulators can play an essential role in providing training opportunities for UAS personnel. Specifically, simulators may allow personnel to repetitively practice tactics and procedures and to meet training proficiency requirements without the limitations of airspace constraints or range availability. UAS are particularly well-suited for simulation training given that UAS vehicle and sensor operators rely on video feeds to perform operations, and DOD and service officials have indicated that current simulators have been used to complete initial training tasks for UAS vehicle and sensor operators.

DOD's current UAS simulators have limited capabilities, however, to enhance training. For example, a recent study performed for DOD found critical deficiencies in each of the UAS training simulators evaluated.[27] In particular, the study found that the military services lacked simulators that were capable of supporting training that is intended to build proficiency in skills required of UAS vehicle and sensor operators and prepare these personnel to conduct UAS combat missions. During our review, we also found several key deficiencies that limit the ability of Air Force and Army simulators to be used for training—including the inability of some simulators to replicate all UAS procedures and to enable the integration of UAS training with other types of aircraft. For example, Air Force officials told us that the Reaper simulator will initially be fielded without weapons-release capabilities, which would enable UAS personnel to replicate the procedures used to attack targets, and this capability will not be available until fiscal year 2011. Similarly, the Army's Shadow Institutional Mission Simulator is not currently capable of replicating system upgrades that are being fielded directly to ongoing combat operations, such as a laser target designator and communications relay equipment. As a result, Shadow unit

---

[26] Department of Defense, *Brigade Combat Team Air-Ground Integration Final Report* (February 2009).

[27] CHI Systems Inc., *UAS Training Simulator Evaluation*, a special report prepared at the request of the United States Special Operations Command, August 2009.

personnel expressed concern that they would be unable to train with these capabilities prior to their deployment.

Air Force and Army simulators are also currently incapable of providing virtual, integrated training opportunities between manned and unmanned aircraft because of interoperability and information security concerns. For example, the Air Force's Predator and Reaper simulators are not interoperable with the Air Force's Distributed Mission Operations Network,[28] which creates a virtual training network for Air Force aviation assets. Officials told us that the Predator and Reaper simulators do not meet Air Force information security requirements for the Distributed Mission Operations Network, which precludes these simulators from participating in virtual integrated training exercises. Similarly, the Army's Shadow Institutional Mission Simulator is not fully interoperable with the Army's manned aviation simulator (the Aviation Combined Arms Tactical Trainer) because of differences in the two simulators' software. According to Army officials, the lack of interoperability of the two simulators detracts from the training value that UAS personnel would receive by performing virtual integrated training with other types of Army aviation assets.

Moreover, the Air Force and the Army have not fully developed comprehensive plans that address long-term UAS simulator requirements and associated funding needs. The Air Force, for example, has not finalized plans to address its UAS simulator goals. Some goals established within the Air Force's UAS *Flight Plan*, such as the development of high-fidelity simulators, are expected to be completed in fiscal year 2010. However, we found that other goals are not linked with the Air Force's funding plans. For example, while officials recognize the training benefit of connecting the Predator and Reaper simulators to the Distributed Mission Operations Network, the Air Force has not identified funds within its future funding plans for this initiative. The Army has not fully defined the number and type of simulators that its active component forces require to meet the training needs of personnel who operate the Shadow and ERMP UAS or the resources needed to acquire these systems. Army officials told us that steps to determine simulator needs are ongoing. Specifically, the Army has commissioned the Army Research Institute to complete a simulator requirements study by October 2010 and it has

---

[28] The Air Force's Distributed Mission Operations Network provides a persistent and secure connection for combat Air Force simulators to perform virtual training exercises.

developed an initial UAS simulation strategy. In contrast, the Army National Guard has begun to acquire a simulator to train soldiers who operate the Guard's Shadow UAS based on the results of a study it completed in 2007 to validate its simulator needs.

DOD Lacks a Comprehensive, Results-Oriented Strategy to Resolve UAS Training Challenges

DOD has identified several challenges that affect service and joint UAS training and has commenced several initiatives intended to address them, but DOD has not developed a comprehensive, results-oriented strategy to prioritize and synchronize these initiatives. A leading practice derived from principles established under the Government Performance and Results Act of 1993[29] is that in order to improve the management of federal agencies, it is important that agencies develop comprehensive strategies to address management challenges that threaten their ability to meet long-term goals. We have previously reported that these types of strategies should contain results-oriented goals, performance measures, and expectations with clear linkages to organizational, unit, and individual performance goals to promote accountability and should also be clearly linked to DOD's key resource decisions.[30]

To address UAS training challenges, DOD has launched a number of initiatives to identify requirements for UAS access to national airspace, to identify available training airspace at current and proposed UAS operating locations, to improve joint training opportunities for ground units and UAS personnel, and to recommend effective training methods and UAS simulator equipment, and these initiatives are at various stages of implementation. Table 3 provides a summary of select DOD organizations and initiatives that are intended to address UAS training challenges.

---

[29] Pub. L. No. 103-62 (1993).

[30] See, for example, GAO-03-192SP, GAO-03-293SP, GAO-07-1072, and GAO-09-175.

Table 3: DOD Organizations and Initiatives Addressing UAS Training Challenges

Lead DOD organizations	Description of initiative	Purpose
U.S. Joint Forces Command - Joint UAS Center of Excellence	National airspace system capabilities-based assessment	Outline requirements for national airspace system access, associated gaps, and potential solutions
U.S. Joint Forces Command - Joint UAS Center of Excellence	Joint UAS minimum training standards	Implement by October 2011 minimum UAS crewmember training tasks to facilitate national airspace system access
U.S. Joint Forces Command - Joint UAS Center of Excellence	UAS integration at predeployment training centers	Provide near-term actionable measures to improve UAS integration at service and joint training centers
U.S. Joint Forces Command - Joint UAS Center of Excellence	UAS training improvement project	Develop a series of documents that a predeployment training center or a unit can use to plan, execute, and assess UAS training events
Office of the Secretary of Defense – Acquisition, Technology, and Logistics UAS Task Force	Civil airspace integration planning and technology development	Review and assess operational requirements, identify acquisition solutions, and recommend training and policy changes necessary to fully integrate UAS into the national airspace system to support DOD requirements
Office of the Secretary of Defense – Personnel and Readiness	UAS training and airspace access study	Complete steps, including documenting UAS training requirements, establishing standard criteria for UAS basing decisions, and identifying supporting training infrastructure requirements
Office of the Secretary of Defense – Personnel and Readiness and U.S. Joint Forces Command - Joint UAS Center of Excellence	UAS surrogate aircraft	Provide manned aircraft equipped with sensor packages to training centers to replicate Predator and Reaper UAS capabilities
Military services and U.S. Special Operations Command	UAS simulation studies	Analyze UAS crewmember missions and training requirements and recommend training methods and equipment to sustain training

Source: GAO analysis of DOD documents.

At the time of our review, DOD's initiatives to improve UAS training were at varying stages of implementation. For example, the Office of the Secretary of Defense's effort to identify UAS airspace and training range requirements was established in October 2008 by the Under Secretary of Defense for Personnel and Readiness. Officials told us that as of January 2010, the team had completed initial meetings and data collection with military service and combatant command officials. As a result of these initial steps, the team has identified specific actions that DOD should take to improve UAS training and airspace access, which include documenting UAS training requirements, establishing criteria for UAS basing decisions, and identifying supporting training infrastructure needs. Further, the Joint UAS Center of Excellence initiated an effort to analyze UAS integration at predeployment training centers in March 2009, and according to officials, they have collected data on UAS training at the National Training Center at Fort Irwin, California, and the Marine Corps Air Ground Combat Center, Twentynine Palms, California. We have previously reported that the Office of the Secretary of Defense's UAS Task Force, established in October 2007, is addressing civil airspace integration planning and technology development, among other issues.[31]

Although many defense organizations are responsible for implementing initiatives to resolve UAS training challenges and to increase UAS access to the national airspace system, DOD has not developed a comprehensive plan to prioritize and synchronize these initiatives to ensure that compatible goals and outcomes are achieved with milestones to track progress. Officials with the Office of the Secretary of Defense who are identifying the amount of DOD-managed airspace at planned UAS operating locations told us that one of their first efforts was to determine whether DOD had developed a comprehensive strategy for UAS training, but that they found that no such strategy existed. These officials also stated that while they intended to complete efforts to improve UAS training and airspace access within 18 months, they had not established specific milestones to measure progress or identified the resources required to achieve this goal. Absent an integrated, results-oriented plan to address the challenges in a comprehensive manner, DOD will not have a sound basis for prioritizing available resources, and it cannot be assured that the initiatives it has under way will fully address limitations in Air Force and Army training approaches.

---

[31] GAO-09-175.

DOD Has Not Fully Incorporated Knowledge Gained from Ongoing UAS Operations in Key Publications

Battlefield commanders and units have increased the operational experience with UAS and have used these assets in innovative ways, underscoring the need for complete and updated UAS publications. We identified several factors that create challenges to incorporating new knowledge regarding UAS practices and capabilities into formal publications in a comprehensive and timely way.

UAS Publications Have Not Been Fully Updated to Include Information to Assist a Range of Stakeholders

DOD components have produced several UAS publications, including service doctrine; multiservice and service-specific tactics, techniques, and procedures; and a joint concept of operations, which are intended to provide military personnel with information on the use of these systems, to address interoperability gaps, and to facilitate the coordination of joint military operations. These publications serve as the foundation for training programs and provide the fundamentals to assist military planners and operators to integrate military capabilities into joint operations. For UAS operations, such stakeholders include both manned and unmanned aircraft operators, military planners in joint operations, and ground units that request UAS assets. Because military personnel involved in joint operations may request or employ assets that belong to another service, they need comprehensive information on the capabilities and practices for all of DOD's UAS. However, many of DOD's existing UAS publications have been developed through service-specific processes and focus on a single service's practices and UAS, and they contain limited information on the capabilities that the other services' UAS could provide in joint operations. This information would assist military personnel at the operational and tactical levels of command to plan for the optimal use of UAS in joint operations and determine the best fit between available UAS capabilities and mission needs. Furthermore, military personnel who are responsible for the effective integration of UAS with other aviation assets in joint operations, such as air liaison officers and joint aircraft controllers, require knowledge beyond a single service's UAS assets and their tactics, techniques, and procedures. To effectively integrate UAS, these service personnel require information that crosses service boundaries, including capabilities, employment considerations, and service employment procedures for all UAS that participate in joint operations.

An internal DOD review of existing key UAS publications conducted in 2009 also found that most of these documents are technical operator manuals with limited guidance to assist military planners and ground units on the employment of UAS in joint operations. For example, the review suggests that military planners and personnel who request the use of UAS assets require additional guidance that links UAS performance capabilities to specific mission areas so that there is a clear understanding of which UAS offer the optimal desired effects. Additionally, these stakeholders also require comprehensive information on UAS planning factors and the appropriate procedures for UAS operators to assist with mission planning.

DOD Has Processes to Capture Knowledge Gained from Ongoing Operations, but Key UAS Publications Do Not Contain Timely Information

In addition, many key publications do not contain timely information. DOD officials told us that existing publications are due for revision given the rapidly expanding capabilities of UAS and the utilization of these assets in joint operations. As a result, information on UAS practices and capabilities described in these publications is no longer current. For example, DOD's multiservice tactics, techniques, and procedures manual for the tactical employment of UAS was last updated in August 2006. According to officials with whom we spoke, the document does not contain detailed information on UAS operations in new mission areas, such as communication relay, fires, convoy support, and irregular warfare.[32] Although DOD components have established milestones to revise UAS publications, in some cases, these efforts have not been successful. For example, the Air Force has canceled conferences that were scheduled to occur in prior fiscal years that were intended to revise the tactics, techniques, and procedures manuals for the Predator UAS because, according to officials, key personnel were supporting overseas operations and were therefore unavailable to participate in the process. As a result, these publications have not been formally updated since 2006, and Air Force officials acknowledged to us that these manuals do not reflect current tactics and techniques. While past attempts to revise these publications have been unsuccessful, the Air Force has scheduled another conference in 2010 to revise the Predator publications.

Documenting timely information on the use of UAS in ongoing joint operations is important because commanders and units are increasing their operational experience with these new weapon systems. As a result, military personnel have often developed and used new approaches to

---

[32] According to officials, DOD's *Multi-Service Tactics, Techniques, and Procedures for the Tactical Employment of Unmanned Aircraft Systems* publication is currently being revised with a planned issuance date in August 2010.

employ UAS, which may differ or build upon approaches outlined in existing publications. For example, according to officials, the use of UAS in ongoing operations has contributed to the development of new tactics for the employment of UAS in counterinsurgency operations—information that has not previously been included in DOD's publications. Officials told us that although publications have not been formally updated, some units, such as Air Force UAS squadrons, maintain draft publications that describe current tactics, techniques, and procedures that are being used in ongoing operations. However, these officials acknowledged to us that while UAS unit personnel have access to these draft documents, other stakeholders, such as military planners and manned aircraft operators, do not have access to the new information contained in the draft publications.

In the absence of updated publications, DOD components have captured lessons learned and developed ad hoc reference materials that contain updated information on UAS capabilities to use in training exercises and during joint operations. For example, the military services and U.S. Joint Forces Command's Joint UAS Center of Excellence maintain Web sites that post lessons learned from recent UAS operations. In addition, warfighter unit personnel with whom we met provided us with several examples of reference materials that were produced to fill voids in published information on current UAS practices. Although this approach assists with documenting new knowledge during the time between publication updates, the use of lessons learned and reference materials as substitutes for timely publications can create challenges in the long term. Namely, these materials may not be widely distributed within DOD, and the quality of the information they contain has not been validated since these materials have not been formally vetted within the normal publication development and review process.

Personnel Availability and Service Coordination Have Limited Development of Comprehensive and Timely Publications

Several factors create challenges to incorporating new knowledge about UAS practices and capabilities into formal publications in a comprehensive and timely way. Because the military services, in some cases, have rapidly accelerated the deployment of UAS capabilities to support ongoing contingency operations, there has been a corresponding increase in new knowledge on the employment of UAS in joint operations. This creates a challenge in incorporating new knowledge and maintaining current information within UAS publications through the normal publication review process. Military service officials noted that the pace of ongoing operations for UAS subject matter experts has also limited the amount of time that key personnel have been available to revise publications. As one example, Air Force officials told us that the subject matter experts who are normally responsible for documenting new tactics, techniques, and procedures within formal manuals for the service's Predator and Reaper UAS are the same service personnel who operate these UAS in ongoing operations. Because of the rapid expansion of the number of Air Force UAS supporting operations, the Air Force has not had enough personnel with critical knowledge on the use of these assets to participate in efforts to update its formal UAS publications. Officials told us that conferences scheduled in previous years intended to update the Predator UAS publications and to develop initial publications for the Reaper UAS were postponed because key personnel were supporting operations and were therefore unavailable to attend the conferences. In 2008, the Air Force established a new squadron at the Air Force Weapons School to develop tactical experts for the service's UAS. According to officials, personnel within the squadron will play a key role in conferences scheduled in fiscal year 2010 that are intended to revise the tactics, techniques, and procedures manuals for both the Predator and Reaper UAS.

We recognize that the pace of operations has strained the availability of key subject matter experts to document timely information in UAS publications, but the military services have not, in some cases, assigned personnel to positions that are responsible for UAS publication development. For example, in 2006, the Air Force established the 561st Joint Tactics Squadron on Nellis Air Force Base, comprising multiservice personnel, with the primary mission to provide timely development and update of tactics, techniques, and procedures publications. However, the squadron did not have UAS subject matter experts on staff who would be responsible for finalizing UAS publications and documenting procedures for the integration of UAS in combat operations, such as in the areas of airspace management and fire support coordination. Squadron officials told us that as of August 2009, the Air Force had not filled its UAS expert positions because of personnel shortfalls throughout the UAS community and the Army had not filled its positions despite agreements between Army and Air Force leadership to do so. According to officials, the lack of these experts also limits the squadron's ability to collect and validate emerging UAS tactics and to disseminate these emerging tactics to warfighters who are preparing to deploy for overseas contingency operations.

Additionally, while a DOD directive[33] makes the services responsible for participating with one another to develop publications for those UAS that are common among the services, they have not yet done so. To their credit, the Army and the Air Force completed a concept in June 2009, which presents a common vision for the services to provide theater-capable, multirole UAS to support a joint force commander across the entire spectrum of military operations. The Army and Air Force view this concept as the first step to improving service-centric UAS procedures, and among other tasks, the services intend to update joint doctrine and tactics, techniques, and procedures for multirole UAS capabilities. However, we found that in several instances, the military services worked independently to develop publications for common UAS and did not maximize opportunities to share knowledge and work collaboratively. The lack of collaboration during the development of publications can limit the sharing of lessons learned and best practices that have been established through the use of UAS in operations. For example:

- In 2009, the Air Force developed the first tactics, techniques, and procedures manual for the Global Hawk UAS, but did not collaborate with the Navy on the process to develop this publication. The Navy is using a similar unmanned aircraft for its Broad Area Maritime Surveillance and has begun operating a version of this UAS to support ongoing operations.
- At the time of our work, the Marine Corps was finalizing its tactical manual for the Shadow UAS, which the service began to deploy in fiscal year 2008. However, the Marine Corps had limited collaboration with the Army in the development of this publication, despite the fact that Army ground units have considerable operational experience employing the Shadow UAS system and have been operating it since 2002.[34]
- We were told that the Air Force did not plan to invite the Army to participate in the process scheduled for 2010 to update the Predator UAS tactics manuals. In 2009, the Army began to deploy an initial version of the ERMP UAS, which is similar in design and performance to the Predator.

---

[33] Department of Defense Directive 5100.1, *Functions of the Department of Defense and Its Major Components* (certified current as of Nov. 21, 2003).

[34] For example, we were told that Army representation in this process was provided by a U.S. Joint Forces Command official.

The lack of comprehensive and timely publications that are written for a range of stakeholders limits the quality of information that is available to serve as the foundation for effective joint training programs and to assist military planners and operators in integrating UAS on the battlefield.

Conclusions

Warfighter demand for UAS has fueled a dramatic growth in DOD's programs and the military services have had success providing assets to military forces supporting ongoing operations. However, the rapid fielding of new systems and the considerable expansion of existing Air Force and Army programs has posed challenges for military planners to fully account for UAS support elements, such as developing comprehensive plans that account for the personnel and facilities needed to operate and sustain UAS programs and ensure the communications infrastructure that is necessary to control UAS operations. While the Air Force and the Army have implemented various actions to address UAS support elements, these actions in many cases have not been guided by a rigorous analysis of the requirements to support UAS programs in the long term or the development of plans that identify milestones for completing actions and synchronize the resources needed for implementation. In the absence of plans that fully account for support elements and related costs, DOD cannot be reasonably assured that Air Force and Army approaches will provide the level of support necessary for current and projected increases in UAS inventories. Moreover, the lack of comprehensive plans limits the ability of decision makers to evaluate the total resources needed to support UAS programs and to make informed future investment decisions. Furthermore, the challenges regarding UAS training may be difficult to resolve unless DOD develops a comprehensive and integrated strategy to prioritize and synchronize the initiatives it has under way to address limitations in Air Force and Army training. Lastly, without assigning personnel or taking steps to coordinate efforts to update and develop UAS publications, information in UAS publications will not be comprehensive and therefore will not include new knowledge on UAS practices and capabilities. This has the potential to limit the quality of information that is available to serve as the foundation for effective joint training programs and to assist military planners and operators in integrating UAS on the battlefield.

Recommendations for Executive Action

We recommend that the Secretary of Defense take the following five actions:

To ensure that UAS inventories are fully supported in the long term, we recommend that the Secretary of Defense direct the Secretary of the Air Force and the Secretary of the Army, in coordination with the Under Secretary of Defense for Acquisition, Technology and Logistics, to conduct comprehensive planning as part of the decision-making process to field new systems or to further expand existing capabilities to account for factors necessary to operate and sustain these programs. At a minimum, this planning should be based on a rigorous analysis of the personnel and facilities needed to operate and sustain UAS and include the development of detailed action plans that identify milestones for tracking progress and synchronize funding and personnel.

To ensure that the Air Force can address the near-term risk of disruption to the communications infrastructure network used to control UAS missions, we recommend that the Secretary of Defense direct the Secretary of the Air Force to establish a milestone for finalizing a near-term plan to provide for the continuity of UAS operations that can be rapidly implemented in the event of a disruption and is based on a detailed analysis of available options.

To ensure that DOD can comprehensively resolve challenges that affect the ability of the Air Force and the Army to train personnel for UAS operations, we recommend that the Secretary of Defense direct the Under Secretary of Defense for Personnel and Readiness, in coordination with the military services and other organizations as appropriate, to develop a results-oriented training strategy that provides detailed information on the steps that DOD will take to

- identify and address the effects of competition and airspace restrictions on UAS training,
- increase the opportunities that Army ground units and Air Force UAS personnel have to train together in a joint environment,
- maximize the use of available assets in training exercises, and
- upgrade UAS simulation capabilities to enhance training.

At a minimum, the strategy should describe overarching goals, the priority and interrelationships among initiatives, progress made to date, milestones for achieving goals, and the resources required to accomplish the strategy's goals.

To help ensure that all stakeholders, including unmanned aircraft operators, military planners, and ground units, have comprehensive and timely information on UAS practices and capabilities, we recommend that the Secretary of Defense direct the Secretary of the Air Force and the Secretary of the Army to assign personnel to update key UAS publications. We also recommend that the Secretary of Defense direct the Secretary of the Air Force, the Secretary of the Army, and the Secretary of the Navy to take steps to coordinate the efforts to develop publications for those UAS where there is commonality among the services.

Agency Comments and Our Evaluation

In written comments on a draft of this report, DOD concurred with four recommendations and partially concurred with one recommendation. DOD's comments are reprinted in appendix II. DOD also provided technical comments, which we incorporated into the report as appropriate.

DOD concurred with our recommendation to direct the Secretary of the Air Force and the Secretary of the Army, in coordination with the Under Secretary of Defense for Acquisition, Technology and Logistics, to conduct comprehensive planning as part of the decision-making process to field new systems or to further expand existing capabilities to account for factors necessary to operate and sustain these programs that at a minimum, is based on a rigorous analysis of the personnel and facilities needed to operate and sustain UAS and include the development of detailed action plans that identify milestones for tracking progress and synchronize funding and personnel. DOD stated that the department conducts ongoing analysis to determine personnel requirements, necessary capabilities for emerging and maturing missions, basing, and training requirements as part of the military services' processes for fielding new systems and expanding existing capabilities and that this planning is based on internal studies as well as rigorous computer modeling, which provides detailed projections of personnel requirements based on anticipated growth and training capacity. DOD further stated that these plans take into account factors that are necessary to operate and sustain UAS, which are applied in order to synchronize funding and personnel. DOD also noted that some planning factors are variable over time and are regularly reassessed in order to validate plans or drive necessary changes. As discussed in the report, the Air Force and the Army are conducting analyses of factors, such as personnel and facilities, which are required to operate and sustain current and projected UAS force levels. However, although the services are requesting funds, they have not finalized ongoing analyses or fully developed plans that specify the actions and resources required to supply the personnel and facilities that are needed to support these inventories in the long term. Therefore, we reiterate our recommendation that as DOD makes decisions to further expand UAS inventories, it needs to ensure that the Air Force and the Army conduct extensive planning, to include performing the necessary analyses for these factors, so that decision makers have complete information on total program costs and assurances that weapon system programs can be fully supported.

DOD concurred with our recommendation to direct the Secretary of the Air Force to establish a milestone for finalizing a near-term plan to provide for the continuity of operations that can be rapidly implemented in the event of a disruption to the communications infrastructure network used to control UAS missions that is based on a detailed analysis of available options. DOD stated the Air Force is conducting a site selection process for identifying a second satellite relay location and that until the alternate site has been selected and funding secured, the Air Force has mitigated risk of communication disruption with a plan for acquiring and positioning backup equipment for the existing satellite relay site. We state in the report that at the time of our review, the Air Force had not conducted a detailed analysis of available options, such as repositioning backup equipment, to determine the extent to which they would provide for the continuity of UAS operations and it had not established a specific milestone to formalize a plan that could be implemented quickly in the event of a disruption. We are encouraged by DOD's statement that the Air Force has since developed a continuity plan. Although we did not have the opportunity to review the plan's contents, we would expect that it is based on a detailed analysis of the equipment that is required to provide a redundant communications capability at the existing satellite relay site and that it includes specific milestones for acquiring and positioning new equipment in the near term.

DOD concurred with our recommendation to direct the Under Secretary of Defense for Personnel and Readiness, in coordination with the military services and other organizations as appropriate, to develop a results-oriented training strategy that provides detailed information on the steps that DOD will take to identify and address the effects of competition and airspace restrictions on UAS training; increase the opportunities that Army ground units and Air Force UAS personnel have to train together in a joint environment; maximize the use of available assets in training exercises; and upgrade UAS simulation capabilities to enhance training. This strategy should, at a minimum, describe overarching goals, the priority and interrelationships among initiatives, progress made to date, milestones for achieving goals, and the resources required to accomplish the strategy's goals. DOD stated that the office of the Under Secretary of Defense for

Personnel and Readiness has work under way to address this recommendation and that organizations, including the offices of the Under Secretary of Defense for Personnel and Readiness and the Under Secretary of Defense for Acquisition, Technology and Logistics, the Joint UAS Center of Excellence, and the military services, are participating on a team to facilitate identifying UAS training requirements and develop a concept of operations for UAS training. DOD further stated that upon completion of the concept, the department will develop and implement a mission readiness road map and investment strategy.

DOD partially concurred with our recommendation to direct the Secretary of the Air Force and the Secretary of the Army to assign personnel to update key UAS publications. DOD stated that military personnel are updating regulations that govern training, certification, and operational guidance for UAS personnel. DOD also stated that the military services are active participants in the process for updating key joint guidance, such as joint publications and other tactics documents, and that the Office of the Under Secretary of Defense for Acquisition, Technology and Logistics is initiating development of the third edition of the *Unmanned Systems Roadmap* and the Joint UAS Center of Excellence is writing the third version of the *Joint Concept of Operations for Unmanned Aircraft Systems*. DOD further stated that guidance on UAS tactics, techniques, and procedures should be incorporated into joint functional guidance rather than the update of documents that are dedicated only to UAS tactics, techniques, and procedures. We state in our report that DOD components, such as the military services and other defense organizations, have produced several publications, including joint and service doctrinal publications, that describe processes to plan for and integrate UAS into combat operations. We also state in the report that DOD components have produced UAS-specific publications, such as multiservice and platform-specific tactics, techniques, and procedures manuals. However, we identified many cases where DOD's UAS publications did not incorporate updated information needed by military personnel to understand current practices and capabilities, and we found that the military services have not, in some instances, assigned personnel to positions that are responsible for UAS publication development. This has the potential to limit the quality of information that is available to serve as the foundation for effective joint training programs and to assist military planners and operators in integrating UAS on the battlefield. Therefore, we continue to believe that our recommendation has merit.

DOD concurred with our recommendation to direct the Secretary of the Air Force, the Secretary of the Army, and the Secretary of the Navy to take steps to coordinate the efforts to develop publications for those UAS where there is commonality among the services. DOD stated that coordination to develop publications where commonality exists between UAS is occurring. For example, DOD stated that the Army and Air Force *Theater-Capable Unmanned Aircraft Enabling Concept* was approved in February 2009. According to DOD, this document outlines how the two services will increase the interoperability of similar systems, and as a result, planning is under way to identify key publications and incorporate joint concepts. As we note in our report, to their credit, the Air Force and Army concept can serve to improve service-centric UAS procedures. However, we found that in other instances, the military services did not maximize opportunities to share knowledge and work collaboratively in the development of UAS publications where there is commonality among the services, which can limit the sharing of lessons learned and best practices that have been established through the use of UAS in operations. Therefore, we reiterate the need for the military services to coordinate the efforts to develop publications for those UAS where there is commonality among the services.

We are sending copies of this report to the Secretary of Defense, the Secretary of the Air Force, the Secretary of the Army, the Secretary of the Navy, and the Commandant of the Marine Corps. This report also is available at no charge on the GAO Web site at http://www.gao.gov.

If you or your staff have any question about this report, please contact me at (202) 512-9619 or pickups@gao.gov. Contact points for our Offices of Congressional Relations and Public Affairs may be found on the last page of this report. Key contributors to this report are listed in appendix III.

Sharon L. Pickup
Director
Defense Capabilities and Management

Appendix I: Scope and Methodology

To address our objectives, we met with officials from the Office of the Secretary of Defense; the Joint Staff; several unified combatant commands; the Multi-National Forces Iraq; and the Departments of the Air Force, the Army, and the Navy who represent headquarters organizations and tactical units. To determine the extent to which plans were in place to account for the personnel, facilities, and communications infrastructure to support Air Force and Army unmanned aircraft systems (UAS) inventories, we focused primarily on Air Force and Army UAS programs that support ongoing operations. Excluded from this review were programs for small unmanned aircraft. While the military services have acquired more than 6,200 of these aircraft, they generally do not have substantial support requirements. We examined the military services' UAS program and funding plans, Department of Defense (DOD) policies governing the requirements definition and acquisition processes, and data generated by the Joint Capabilities Integration and Development System—the department's principal process for identifying, assessing, and prioritizing joint military capabilities and the process used by acquisition personnel to document a weapon system's life cycle costs (including support costs) to determine whether the associated program is affordable. We analyzed UAS funding requests included in the President's budget requests for fiscal years 2006 through 2010. We compiled data from the Departments of the Air Force, the Army, and the Navy and the DOD-wide procurement, research, development, test and evaluation, military construction, and operation and maintenance budget justification books.[1] We reviewed documents that detail UAS operational concepts and we interviewed officials with the Office of the Secretary of Defense and the military services to determine whether UAS plans account for the services' personnel, facilities, and communication infrastructure needs for these concepts, and to determine any actions taken to update UAS plans to more accurately reflect the costs of further expanding UAS programs. We considered all of the information collected on these planning efforts in light of knowledge gained by the services from operational experiences with the use of UAS in ongoing contingency operations. In examining UAS planning documents, we consulted the Office of Management and Budget's *Capital Programming Guide* and our *Cost Estimating and Assessment*

---

[1] All of the associated costs for UAS programs are not transparent within the budget justification books. We requested supplementary data from the services to provide additional information regarding operation and support costs as well as facility construction or renovation costs.

Appendix I: Scope and Methodology

Guide for instruction on developing cost estimates and plans to manage capital investments.[2]

In determining the extent to which DOD addressed challenges that affect the ability of the Air Force and the Army to train personnel for UAS operations, we visited select military installations and the Army's National Training Center at Fort Irwin, California, and spoke with knowledgeable DOD and military service officials to determine the specific challenges that the Air Force and the Army faced when training service personnel to perform UAS missions in joint operations. Specifically, we spoke with Air Force and Army personnel in UAS units in the United States and in Iraq to determine the training that they were able to perform prior to operating UAS in joint operations through live-fly training and through the use of simulators. We discussed the challenges, if any, that prevented them from performing required training tasks. In identifying Air Force and Army unit personnel to speak with, we selected a nonprobability sample of units that were preparing to deploy for contingency operations or had redeployed from these operations from May 2009 through September 2009. We examined documents and spoke with DOD and military service officials to identify initiatives that have begun to address UAS training challenges. We assessed DOD's efforts to overcome these challenges in light of leading practices derived from principles established under the Government Performance and Results Act of 1993, which are intended to assist federal agencies in addressing management challenges that threaten their ability to meet long-term goals, and key elements of an overarching organizational framework, such as developing results-oriented strategies, as described in our prior work.[3]

To determine the extent to which DOD updated its existing publications that articulate doctrine and tactics, techniques, and procedures to reflect the knowledge gained from using UAS in ongoing operations, we examined joint, multiservice, and service-specific UAS doctrine, tactics, techniques, and procedures, and concept of operations publications. We interviewed DOD and military service officials to determine which organizational entities require information on UAS capabilities and practices. We examined the publications to determine the level of information provided to various organizations and personnel that are responsible for planning for and employing UAS in joint operations. We

---

[2] See Office of Management and Budget, *Capital Programming Guide*, and GAO-09-3SP.

[3] See, for example, GAO-03-192SP, GAO-03-293SP, GAO-07-1072, and GAO-09-175.

Appendix I: Scope and Methodology

also analyzed the publications to determine the degree to which information is provided to the various organizations and personnel that are responsible for planning for and employing UAS in joint operations. Finally, we interviewed DOD and military service officials about the processes used to develop and update publications; any challenges that affect their ability to update key publications; and how new knowledge regarding UAS operations, such as lessons learned and best practices, is captured. We analyzed these processes to determine the level of coordination among the military services to develop UAS publications and the frequency at which documents have been revised.

We conducted this performance audit from October 2008 through March 2010 in accordance with generally accepted government auditing standards. Those standards require that we plan and perform the audit to obtain sufficient, appropriate evidence to provide a reasonable basis for our findings and conclusions based on our audit objectives. We believe that the evidence obtained provides a reasonable basis for our findings and conclusions based on our audit objectives.

We interviewed officials, and where appropriate obtained documentation, at the following locations:

Office of the Secretary of Defense	<ul><li>Office of the Under Secretary of Defense for Acquisition, Technology and Logistics</li><li>Office of the Under Secretary of Defense for Intelligence</li><li>Office of the Under Secretary of Defense for Personnel and Readiness</li><li>Office of the Director, Cost Assessment and Program Evaluation</li></ul>
Department of the Air Force	<ul><li>Office of the Deputy Chief of Staff for Manpower and Personnel</li><li>Office of the Deputy Chief of Staff for Intelligence, Surveillance, and Reconnaissance</li><li>Air Combat Command<ul><li>432nd Wing</li><li>6th Combat Training Squadron</li><li>561st Joint Tactics Squadron</li></ul></li><li>Air Force Central Command<ul><li>609th Combined Air Operations Center</li><li>332nd Expeditionary Operations Group</li></ul></li></ul>
Department of the Army	<ul><li>Office of the Deputy Chief of Staff, G3/5/7</li><li>Army Corps of Engineers</li><li>Army National Guard</li><li>Army Forces Command</li><li>Army Installation Management Command<ul><li>Fort Bragg, North Carolina</li></ul></li></ul>

	• Fort Carson, Colorado
	• Fort Drum, New York
	• Fort Hood, Texas
	• Fort Huachuca, Arizona
	• Fort Irwin, California
	• Fort Lewis, Washington
	• Fort Riley, Kansas
	• Fort Stewart, Georgia
	• Army Materiel Command
	   • Program Executive Office-Aviation, Program Manager UAS
	• Army Research, Development, and Engineering Command
	   • Aviation and Missile Research Development and Engineering Center, Joint Technology Center/Systems Integration Laboratory
	• Army Training and Doctrine Command
	   • Army Aviation Center of Excellence
	• 1st Cavalry Division
	   • 4th Brigade Combat Team
	• 2nd Infantry Division
	   • 3rd Stryker Brigade Combat Team
	• 3rd Infantry Division
	   • 1st Brigade Combat Team
	• 4th Infantry Division
	   • 1st Brigade Combat Team
	   • 3rd Brigade Combat Team
	   • 4th Brigade Combat Team
	• 10th Mountain Division
	• 10th Army Special Forces Group

Department of the Navy	• Research, Development, and Acquisition • Program Executive Office for Unmanned Aviation and Strike Weapons, Persistent Maritime Unmanned Aircraft Systems • Space and Naval Warfare Systems Command • Space and Naval Warfare Systems Center Pacific • Headquarters Marine Corps, Department of Aviation, Weapons Requirements Branch
Other DOD Components	• Multi-National Forces Iraq • Multi-National Corps Iraq • United States Central Command • United States Joint Forces Command • United States Special Operations Command

Appendix II: Comments from the Department of Defense

OFFICE OF THE UNDER SECRETARY OF DEFENSE
3000 DEFENSE PENTAGON
WASHINGTON, DC 20301-3000

ACQUISITION,
TECHNOLOGY
AND LOGISTICS

MAR 15 2010

Ms. Sharon L. Pickup
Director, Defense Capabilities and Management
U.S. Government Accountability Office
441 G Street, N.W.
Washington, DC 20548

Dear Ms. Pickup:

This is the Department of Defense (DoD) response to the GAO draft report, GAO-10-331, "UNMANNED AIRCRAFT SYSTEMS: Comprehensive Planning and a Results-Oriented Training Strategy are Needed to Support Growing Inventories," dated February 4, 2010 (GAO Code 351271).

The DoD partially-concurs with the draft report's fourth recommendation and concurs with recommendation one through three and five. The rational for the DoD's position is enclosed.

The Department appreciates the opportunity to comment on the draft report. For further questions concerning this report, please contact Mr. Edward Wolski, Unmanned Warfare, Edward.Wolski@osd.mil, 703-695-8778.

Sincerely,

David G. Ahern
Director
Portfolio Systems Acquisition

Enclosure:
As stated

Appendix II: Comments from the Department of Defense

GAO DRAFT REPORT DATED FEBRUARY 4, 2010
GAO-10-331(GAO CODE 351271)

"UNMANNED AIRCRAFT SYSTEMS: COMPREHENSIVE
PLANNING AND A RESULTS-ORIENTED TRAINING STRATEGY
ARE NEEDED TO SUPPORT GROWING INVENTORIES"

DEPARTMENT OF DEFENSE COMMENTS
TO THE GAO RECOMMENDATIONS

RECOMMENDATION 1: The GAO recommends that the Secretary of Defense direct the Secretary of the Air Force and the Secretary of the Army, in coordination with the Under Secretary of Defense for Acquisition, Technology, and Logistics, to conduct comprehensive planning as part of the decision making process to field new systems or to further expand existing capabilities to account for factors necessary to operate and sustain these programs. At a minimum, this planning should be based on a rigorous analysis of the persons and facilities needed to operate and sustain UAS and include the development of detailed action plans that identify milestones for tracking progress and synchronize funding and personnel. (Page 39/GAO Draft Report)

DOD RESPONSE: Concur. The Department of Defense (Department) conducts ongoing analysis to determine personnel requirements, necessary capabilities for emerging and maturing missions, basing, and training requirements as part of the Military Service's Title 10-guided process for fielding new systems and expanding existing capabilities. This planning is based on internal studies as well as a rigorous computer modeling which provides detailed projections of personnel requirement based on anticipated growth and training capacity. These plans take into account factors that are necessary to operate and sustain UAS, which are applied in order to synchronize funding and personnel. Some planning factors are variable over time, however, and are regularly reassessed in order to validate plans or drive necessary changes.

RECOMMENDATION 2: The GAO recommends that the Secretary of Defense direct the Secretary of the Air Force to establish a milestone for finalizing a near-term plan to provide for the continuity of UAS operations that can be rapidly implemented in the event of a disruption and is based on a detailed analysis of available options. (Page 40/GAO Draft Report)

DOD RESPONSE: Concur. The Air Force is conducting the site selection process for identifying a second satellite relay location. Until the alternate site has been selected and the MILCON funding secured, the Air Force has mitigated this risk of communication

Attachment
Page 1 of 3

Appendix II: Comments from the Department of Defense

disruption with a plan for acquiring and positioning back-up equipment for the satellite relay sites.

RECOMMENDATION 3: The GAO recommends that the Secretary of Defense direct the Under Secretary of Defense for Personnel and Readiness, in coordination with the Military Services and other organizations as appropriate, to develop a results-oriented training strategy that provides detailed information on the steps that DoD will take to:

- Identify and address the effects of competition and airspace restrictions on UAS training,
- Increase the opportunities that Army ground units and Air Force UAS personnel have to train together in a joint environment
- Maximize the use of available assets in training exercises, and
- Upgrade UAS simulation capabilities to enhance training.

At a minimum, the strategy should describe overarching goals; the priority and inter-relationships among initiatives, progress made to date; milestones for achieving goals; and the resources required to accomplish the strategy's goals. (Page 40/GAO Draft Report)

DOD RESPONSE: Concur. The office of the Under Secretary of Defense for Personnel and Readiness (USD(P&R)) has work underway now to address this recommendation. ODUSD(P&R) is leading a UAS Tiger Team in coordination with the Office of the Under Secretary of Defense for Acquisition, Technology, and Logistics (OUSD(AT&L)) UAS Task Force, the Joint UAS Center of Excellence (JUAS COE) and the Military Services to facilitate identification of training requirements and develop a concept of operations (CONOPS) for UAS continuation training. As part of this effort, the areas identified above will be assessed and incorporated into the CONOPS. Upon completion of the CONOPS a UAS Mission Readiness Roadmap and Investment Strategy will be developed and implemented.

RECOMMENDATION 4: The GAO recommends that the Secretary of Defense direct the Secretary of the Air Force and the Secretary of the Army to assign personnel to update key UAS publications. (Page 41/GAO Draft Report)

DOD RESPONSE: Partially Concur. Military personnel are updating regulations which govern the training, certification, and operational guidance for UAS crews. The Military Services are also active participants in the process for updating key joint guidance such as Joint Publication (JP) 3-30, the 3-09 series of JPs, and JFIRE, and other key tactics documents. UAS guidance on tactics, techniques, and procedures for their employment should be incorporated into joint functional guidance vs. the update of documents that are

Attachment
Page 2 of 3

Appendix II: Comments from the Department of Defense

dedicated only to UAS Tactics Techniques and Procedures. The OUSD(AT&L) UAS Task Force is initiating the third edition of its "*Unmanned Systems Roadmap*" and the JUAS COE is writing the third version of the JUAS CONOPS.

RECOMMENDATION 5: The GAO recommends that the Secretary of Defense direct the Secretary of the Air Force, the Secretary of the Army, and the Secretary of the Navy to take steps to coordinate the efforts to develop publications for those UAS where there is commonality among the services. (Page 41/GAO Draft Report)

DOD RESPONSE: Concur. Coordination to develop publications where commonality exists between UAS is occurring. For example, the Army/Air Force Theater-Capable Unmanned Aircraft Enabling Concept, a product of the Air Force Air Combat Command and the Army Training and Doctrine Command collaboration, was approved by both the Chief of Staff Army and the Chief of Staff Air Force in February 2009. This document outlines how the two services will increase the interoperability of similar systems. As a result, planning is underway to identify key publications and incorporate joint concepts.

Appendix III: GAO Contact and Staff Acknowledgments

GAO Contact	Sharon L. Pickup, (202) 512-9619 or pickups@gao.gov
Acknowledgments	In addition to the contact named above, Patricia Lentini, Assistant Director; Meghan Cameron; Mae Jones; Susan Langley; Ashley Lipton; Greg Marchand; Brian Mateja; Jason Pogacnik; Mike Shaughnessy; and Matthew Ullengren made significant contributions to this report.

GAO's Mission	The Government Accountability Office, the audit, evaluation, and investigative arm of Congress, exists to support Congress in meeting its constitutional responsibilities and to help improve the performance and accountability of the federal government for the American people. GAO examines the use of public funds; evaluates federal programs and policies; and provides analyses, recommendations, and other assistance to help Congress make informed oversight, policy, and funding decisions. GAO's commitment to good government is reflected in its core values of accountability, integrity, and reliability.
Obtaining Copies of GAO Reports and Testimony	The fastest and easiest way to obtain copies of GAO documents at no cost is through GAO's Web site (www.gao.gov). Each weekday afternoon, GAO posts on its Web site newly released reports, testimony, and correspondence. To have GAO e-mail you a list of newly posted products, go to www.gao.gov and select "E-mail Updates."
Order by Phone	The price of each GAO publication reflects GAO's actual cost of production and distribution and depends on the number of pages in the publication and whether the publication is printed in color or black and white. Pricing and ordering information is posted on GAO's Web site, http://www.gao.gov/ordering.htm. Place orders by calling (202) 512-6000, toll free (866) 801-7077, or TDD (202) 512-2537. Orders may be paid for using American Express, Discover Card, MasterCard, Visa, check, or money order. Call for additional information.
To Report Fraud, Waste, and Abuse in Federal Programs	Contact: Web site: www.gao.gov/fraudnet/fraudnet.htm E-mail: fraudnet@gao.gov Automated answering system: (800) 424-5454 or (202) 512-7470
Congressional Relations	Ralph Dawn, Managing Director, dawnr@gao.gov, (202) 512-4400 U.S. Government Accountability Office, 441 G Street NW, Room 7125 Washington, DC 20548
Public Affairs	Chuck Young, Managing Director, youngc1@gao.gov, (202) 512-4800 U.S. Government Accountability Office, 441 G Street NW, Room 7149 Washington, DC 20548

www.ingramcontent.com/pod-product-compliance
Lightning Source LLC
Chambersburg PA
CBHW081617170526
45166CB00009B/3008